WHEAT
The Golden Harvest

WHEAT
The Golden Harvest

Dorothy Hinshaw Patent
Photographs by William Muñoz

DODD, MEAD & COMPANY
New York

ACKNOWLEDGMENTS

The author and photographer wish to thank the crew at Bernice's Bakery for their helpful cooperation, Greg Patent for his time and creativity, and Larry and Susan Hansen for their information and for their reading of the manuscript.

Distributed in Canada by
McClelland and Stewart Limited, Toronto
Printed in Hong Kong by South China Printing Company

1 2 3 4 5 6 7 8 9 10

Library of Congress Cataloging-in-Publication Data

Patent, Dorothy Hinshaw.
  Wheat, the golden harvest.
  Includes index.
  Summary: Text and photographs describe how different varieties of wheat are planted, harvested, and processed into foods.
  1. Wheat—Juvenile literature. 2. Wheat products—Juvenile literature. [1. Wheat]
I. Muñoz, William, ill. II. Title.
SB191.W5P337 1987   633.1'1   86-32801
ISBN 0-396-08781-7

For Greg Patent,
who can make so many delicious treats from wheat

CONTENTS

1

OUR MOST IMPORTANT FOOD

What do you eat for breakfast in the morning? Do you have a bowl of tasty cereal? French toast or pancakes? A muffin? Chances are, you put away some foods like these made with wheat every day.

You may not realize how much wheat you eat. In the store, only some bread is labeled "wheat bread." It is usually brown in color. But white bread is made from wheat, too. Favorite foods such as cakes, pie crusts, dinner rolls, cookies, spaghetti, noodles, and many breakfast cereals are all made from wheat. In fact, each American eats about a hundred pounds of wheat products in a year.

WHAT IS WHEAT?

Wheat is the most important food in the world. It gives more nourishment to more people than any other food. Every month of the year, wheat is har-

A lone farmhouse is surrounded by fields in the rich wheat-growing country of eastern Washington.

vested somewhere in the world. It can be grown almost anywhere except near the frigid poles and in the hot, humid tropics.

Wheat is a kind of grain. And grains are the seeds of some plants belonging to the grass family. They are also called cereals. Rice, barley, oats, rye, and corn are cereal crops, too. The grain, or kernel, of wheat is smaller than your little fingernail. It takes about fifteen thousand wheat kernels to make a pound.

The kernel has an outer covering called the bran. The bran has many thin layers. It makes up about 15 percent of the kernel. Inside the bran, the kernel itself has two parts. Near the bottom is a very important area called the germ, which only makes up about 3 percent of the grain. It will grow into a new plant if

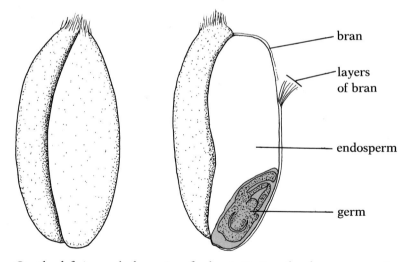

On the left is a whole grain of wheat. Notice the deep groove. On the right, the grain has been cut to show the parts.

Bleached white flour, on the left, is whiter than the unbleached white flour on the right. Both these and whole wheat flour, below, are used for baking breads and pastries.

the seed is put into the ground. The rest of the kernel is called the endosperm. The endosperm nourishes the young plant as it grows.

Whole wheat flour, which is made from the entire wheat grain, is light brown. Most wheat products in

America are made from white flour, in which the bran and germ are removed. White flour is usually bleached. Then it is very white. Unbleached white flour is pale cream in color.

WHERE DOES WHEAT COME FROM?

Many thousands of years ago, people collected the grains of wild wheat for food. It was hard work, for the seeds fell from the plants when they were ripe. They had to be picked up one by one. And once the grain was collected, it was difficult to separate the kernels from the layers surrounding them.

Perhaps ten thousand years ago, the idea of planting wheat seeds came to people somewhere in the Middle East. Then the wheat would grow where they wanted it. They wouldn't have to go out in the fields and look for it anymore.

Now and then, unusual plants that held the ripe seeds or had kernels easy to separate probably sprung up in those ancient fields. Some had bigger kernels than others, too. Such plants must have been valued for their useful traits. Their seeds were planted for the next year's harvest, slowly changing wheat from a wild grass to a valuable cultivated crop. The earliest cultivated wheat that looked different from wild types was probably raised by villagers in Iraq, Turkey, and Iran around 7,500 BC.

Wheat ready for harvest.

Young wheat plants. The one in front at the center is putting up a spike.

HOW WHEAT GROWS

Wheat is much bigger than the grass in lawns. It usually grows to be two to four feet tall. After sprouting many leaves, each plant sends up from twenty to fifty stalks of flowers, called spikes. The flowers don't look like the ones you are used to. Familiar flowers are brightly colored to attract insects which take pollen from one flower to another to pollinate them. But wheat flowers pollinate themselves. They are small and not very noticeable, since they don't need to have insects find them.

After the flowers are pollinated, the seeds develop. The mature spike of wheat seeds is called a

14

A green head of winter wheat, before the kernels have filled out completely.

head. Each head has twenty to fifty kernels of wheat. The plants dry out after they have made seeds, producing vast stretches of golden fields. Then the wheat is harvested.

KINDS OF WHEAT

Pastry is flaky. Spaghetti is slightly chewy. Bread is full of bubbles. How can one crop make such dif-

Soft white wheat, on the left, looks very different from hard red wheat, below. On the right are rolled grains of wheat, which can be cooked like oatmeal.

ferent kinds of delicious things to eat? These three major wheat foods come from different sorts of wheat. Bread is made mostly from hard wheat. Most of the American harvest is hard wheat. It has lots of gluten, a special protein in the endosperm. When bread dough is kneaded, the gluten becomes stretchy. Then when the bread rises, gas bubbles are trapped in the dough, making it light and airy.

Pastry is made from soft wheat. Soft wheat has little gluten. Foods made with soft wheat are flaky and tender. Bits of pie crust don't hold together like bread because they have little gluten. If pastry is made with hard wheat, it is tough, not tender.

Spaghetti, macaroni, and noodles are all kinds of pasta. Pasta is made with durum wheat, a special kind of very hard wheat. If pasta is made with soft wheat flour, the strands stick together and are gummy when cooked. But pasta made from durum wheat stays separate.

Soft wheat is used to make pie crust because it produces flaky, tender pastry.

2

GROWING WHEAT

Wheat is a vital crop in southern Canada and many states from Texas to the Canadian border. While some wheat grows in all parts of the country except Florida and northern New England, wheat does best where summers are not too hot and humid. The top states for wheat production are Kansas, North Dakota, Montana, Nebraska, and Oklahoma.

How can wheat produce so well in such different places as North Dakota, with its long bitter winters and short summers, and Oklahoma, with its short winter and scorching hot summers? The secret lies in the adaptability of wheat and its two major types. Through most of North America, winter wheat thrives. It is planted in the fall and harvested the following summer or early autumn. In the Dakotas, eastern Montana, and Canada, where winters are too severe for winter wheat to survive, spring wheat saves the day. It is usually planted in May and harvested in

August or September. Spring wheat can even be grown in Fairbanks, Alaska.

PLANTING WHEAT

Before seeds are planted, the ground has to be made ready for its new crop. First, the soil is turned to loosen it and to kill young weed plants that have started to grow. The ground can be turned with a plow, or it can be disked—worked by a set of metal disks that cut deeply into the soil. Then the soil must be evened out with a harrow, which has many fine teeth that break up lumps of dirt, leaving a smooth, even surface ready for another year of growth.

A plowed field awaits the spring, when it will be worked some more and then planted.

A disk waiting in the field.

A harrow works a field before planting wheat. On each end, part of the harrow is off the ground so you can see the teeth.

20

Wheat is planted by a cleverly designed machine called a grain drill, which carries out the whole operation at one time. First, a shovel cuts into the soil, making a trench for the seeds. Then a notched wheel allows the seeds to pass one at a time from a storage hopper through a tube that opens just behind the shovel. The seeds drop one by one into the trench right after it has formed. As the grain drill rolls onward, a wheel presses soil down over the newly planted seeds. Now they are ready to grow.

A notched wheel separates the seeds so only one is planted at a time. Behind each tube carrying seeds is one that releases fertilizer to enrich the soil.

21

Overleaf: A grain drill.

Growing Winter Wheat

In the fall, when winter wheat is planted, the ground is usually moist from fall rains. The seed swells with water from the surrounding soil and begins to grow. First it sends out tiny roots that stretch down into the ground. Then a shoot pushes upward through the dark soil into the sunlight and air above. Until the shoot is big enough to gather energy from the light, the growing plant is nourished from the endosperm in the kernel.

Before the cold weather settles in, the winter wheat plant is four to eight inches tall. Some farmers leave the wheat alone, but others let cattle feed on it so that the plants are very short through winter.

During the cold season, the wheat rests quietly under the snow. While it waits, the wheat takes in a special message from the cold weather. The low winter temperature signals the plant to grow up and go to seed when it gets warm again. If winter wheat is planted at a different time of year and doesn't go through a period of cold temperatures, it won't produce spikes later on.

When spring arrives, the wheat surges with growth. By early summer, the spikes shoot up, the flowers are pollinated, and the kernels form. While the plant was taking in energy from the sunlight, nutrients were stored in the leaves. Now these nutrients flow to the growing kernels, making them plump with healthful nourishment.

Young wheat plants.

Winter wheat peeking through the snow.

A field of stubble in winter. No snow spells trouble for the wheat farmer, because the moisture in the soil can dry out. In planted fields, freezing and thawing of the soil can damage young plants.

Wheat ready for harvest in Oregon, with Mt. Hood in the background.

Once the kernels are ripe, the wheat turns golden and the seeds dry out. It is important that the grain be dry enough before harvest. The kernels should have only about 12 percent to 13 percent water. Grain with more than that can become moldy or otherwise spoil in storage.

TAKING CARE OF THE CROP

Most wheat farmers depend on nature to water their crops. Sprinkling the largest wheat fields would be impossible. One reason wheat is so widely grown is that it can survive and produce a crop on very little water. Even so, fields in dry areas with only twelve to fifteen inches of rain each year yield only a fifth as much as if they were watered, or irrigated.

In dry states such as Montana, some wheat fields are irrigated by huge, moving sprinklers.

Green fields of wheat and golden fallow fields are separated by a plowed firebreak.

In very dry areas, such as parts of Colorado, Montana, and Washington, unirrigated fields produce wheat only every two years. After harvest, the field is left through the winter with the remains of the cut plants, or stubble. In the spring, the soil is turned, but no wheat is planted, leaving the field fallow. Since no plants are growing, vital moisture and nutrients can build up to help nourish the crop the following year.

Wheat being planted by a no-till drill. Behind the drill is a tank of ammonia fertilizer that is applied to the field as it is planted.

Another way to conserve moisture is called no-till. With no-till, the soil isn't turned to stop weeds. Instead, chemicals called herbicides that kill weeds are sprayed on the ground. No-till farming also helps slow wind erosion. When the bare, dry soil is exposed to the wind, tons of it are blown away. This removes nutrients and makes the soil less fertile for growing crops. Soil carried by the wind can also damage young wheat plants and cover them with dirt so they don't grow well.

A row of wheat grows between rows of stubble in no-till farming.

Windblown soil on the snow is a result of wind erosion.

These young wheat plants have been damaged by sharp particles of blowing soil. Notice that the soil has also covered up the bottoms of the plants.

Too much water can also wash away the topsoil and damage the fields. If water erosion isn't controlled, gulleys that carry both soil and water away from the fields might form, making the land less productive and harder to work with machinery.

Wheat needs fertilizer to thrive. Often, fertilizer is added to the soil as the seeds are planted. It can also be sprayed on at a later time. Even with tilled ground, some weeds sprout along with the wheat, so herbicides are sprayed on the fields. Wheat can also be damaged by insect pests and by several diseases. Again, wheat farmers use chemicals to control the insects and the diseases.

Water can damage fields by carrying away precious topsoil. This shallow gulley will become deeper and deeper each year if the farmer doesn't work to control the erosion of his fields.

When weeds grow with the wheat, they take precious nutrients, water, and space away from the crop.

34

In a bad year, grasshoppers can cause considerable damage to a wheat crop.

Many people worry about the chemicals that get into our water supply and into living things from fertilizing and controlling weeds, pests, and diseases of wheat and other crops. No one knows for sure just what effects these chemicals will have. Scientists are studying the problem, however, and hope to find ways to solve it.

A combine harvests wheat.

HARVESTING WHEAT

In America, the wheat harvest begins in May in Texas and Oklahoma and travels steadily northward as the crop ripens. Giant machines called combines do the harvesting. The combine gets its name from its work. It combines the process of cutting the wheat, called reaping, with separating the grain from the husks and stems, called threshing.

The huge combines quickly get the job done. A combine, operated by just one person sitting at the

36

controls in its cool, air-conditioned cab, can do work that once took twelve sweating men laboring through the heat with a dozen horses, mules, or oxen. In the old days, it took three days to cut and thresh just one acre of wheat. A combine can roll through an acre, accomplishing the same work, in only six minutes!

At the front of the combine is a long reel with metal strips. The strips slice into the rows of wheat and press the plants against a cutter bar at the front edge of the combine. The cut wheat then rides a conveyor belt into the complex workings of the machine. There, the kernels are threshed—separated from the straw (dried stems and leaves) and from the chaff, the other parts of the head. The straw and chaff are dumped

In order for the combine to reap the grain, plants must be upright. Sometimes a strong wind, rain, or hail will knock the wheat down so that it cannot be harvested.

The rotating bars on the front of the combine press the wheat plants against the cutter bar.

The combine releases the straw and chaff onto the harvested field behind it as it works.

38

back onto the harvested field, and the grain is poured into a tank on the combine.

When the combine tank fills up, the grain is transferred to a truck. The truck takes it to a big building called a grain elevator for storage. At the elevator, the grain is checked over for quality. The better the wheat, the more money the farmer gets for it.

Wheat is measured by the bushel. A bushel of wheat weighs about sixty pounds. It can be processed into forty-two pounds of white flour, which will make sixty-six loaves of bread.

Grain is loaded onto a truck from a combine.

Grain elevators in the sunset.

Grains of wheat in the truck. The combine has done all the complicated work of reaping and threshing the wheat in a matter of minutes. Only a few bits of material other than the grain are left.

3

AFTER THE HARVEST

Grains are especially useful crops because they can be stored for such a long time without spoiling. It may be months after harvest before the grain is processed into food. When wheat is sold, it is shipped from the elevators by truck, train, or river barge.

FROM WHEAT TO FLOUR

White flour is the most common wheat product in America. It keeps better than whole wheat flour. The germ, which is removed in making white flour, has lots of oil in it. Because of the oil, whole wheat flour spoils faster than white flour. White flour also makes softer, fluffier bread and flakier pastry than whole wheat flour. But because the bran and germ have been removed, white flour is not as nutritious as whole wheat. Whole wheat flour has more B vi-

42

A huge grain elevator in Hutchinson, Kansas, dwarfs the railroad cars waiting to take on a load of wheat. Hutchinson has the biggest grain elevator in the world.

Wheat being loaded into a railroad car.

tamins, fiber, protein, and more of many minerals than white flour.

At the mill, the wheat grains need to be cleaned so that no stones, weed seeds, or bits of metal are mixed in. The wheat is run through sorting disks that only allow material the size of wheat grains to pass. A giant magnet removes metal pieces. Finally, the wheat is bathed in a special tub so that lighter material floats to the top and stones fall to the bottom.

Making white flour is a complicated process. Unfortunately for millers, wheat kernels have a deep crease along one side. This makes removing all of the bran difficult. The grains are fed between heavy rollers that break them into pieces. The pieces then pass through screens that sort them for size. Over and over, the pieces drop through sets of rollers and screens, getting smaller and whiter each time. As the bran is worn off the grains, it is blown away by fans and collected separately.

Eventually, the clean bits of endosperm are crushed very finely by rollers to become flour. The flour then flows through a fine sieve to make sure it is evenly ground and is bagged.

BRAN AND GERM

Wheat germ and bran are used for food, too. The germ contains lots of important nutrients. It is offered flaked and sometimes toasted for adding to breakfast

Whole wheat flour can be made using a home grain mill, which is quite simple in structure.

The grains of wheat are fed through a hole to the grinding stones.

The two stones rotate against each other, grinding the grain into flour. The flour pours out through the hole in the lower stone.

45

cereal, homemade bread, and other foods. Many breakfast cereals have extra bran to add healthful fiber. The bran and germ, as well as unmilled wheat kernels, are sometimes used for animal feed.

OTHER WHEAT PRODUCTS

Not all wheat becomes white or whole wheat flour. Large quantities are turned into popular breakfast cereals. Durum wheat is made into semolina, which is coarser than flour. To make favorite foods such as

Pasta comes in many shapes and sizes.

46

There are many different wheat products. At the upper right is semolina, and below that is wheat germ. On the lower left is bulgur, with bran above.

spaghetti and elbow macaroni, the semolina is mixed with water. Noodles are made by mixing the semolina with eggs.

Wheat kernels may also be cracked, steamed, and dried to make bulgur. Bulgur is very popular in the Middle East. It is cooked with water, much like rice. Grains of wheat are also rolled so that they can be cooked for breakfast like oats.

4

MAKING BREAD

Bread, in one form or another, is eaten by people around the world. Some kinds, like Swedish lefse and Jewish matzoth, are unleavened. They are flat and often heavy because they are made without yeast. Our familiar tasty loaves, dotted with airy bubbles, owe their delightful texture to yeast. Plain French bread is made with just yeast, flour, water, and a little salt. Other kinds of bread have added ingredients, such as eggs, milk, sugar, oil, wheat germ, or potatoes, which affect their taste and texture.

Different grains, such as rye and oats, are also often used for bread. Some wheat flour is added in making these breads, however, because other grains have little or no gluten. Even when made partly with wheat flour, rye bread tends to be dense and heavy. The gas bubbles are not trapped in the dough as well since less gluten is present.

The bread labeled "wheat bread" in stores has

some whole wheat flour in it along with the white flour. Most of the brown color, however, usually comes from food coloring, for such bread often has very little whole wheat flour. Because the bran and germ do not have gluten and are usually coarse, bread made from 100 percent whole wheat flour is not as light and fluffy as white bread or store-bought "wheat bread."

YEAST, THE BAKER'S HELPER

Yeast is made of tiny living cells. These cells use the starch and sugar in the flour to get energy to grow and reproduce. In the process, the gas carbon dioxide is released.

Nowadays, yeast is usually sold dry. It will keep that way for months. The cells in the dry yeast are resting. To make bread, the dry yeast is mixed with water and a little sugar. The yeast cells swell with

A plump loaf of tasty egg bread.

On the left, yeast is added to water that has a little sugar in it. On the right, the dissolved yeast has bubbled up, showing that it is alive and ready to be used to make bread.

water and begin to grow and divide. When the yeast bubbles up, the baker knows that the cells are alive.

Making the Dough

After the yeast has dissolved, it is added along with more liquid and any other ingredients to the flour. At first, the bread mixture is lumpy. It doesn't look very appetizing. Then it is kneaded. The kneading can be done with a machine or by hand. The dough is pushed this way and that, which "activates" the gluten, making it become stretchy. Within a few minutes, the lumpy mass of dough becomes smooth and sleek.

After the dough is kneaded, it is covered and left to rise. During the rising, the bubbles of carbon dioxide released by the yeast are caught in the stretchy dough so that it puffs up.

After it has risen to about twice its original size, the dough is punched down to get out the bubbles. Then it is shaped into loaves and left to rise again. As the yeast works on the dough, it improves the flavor through the chemical changes it makes as well as making the bread light and fluffy.

Once the loaves have risen, they are baked. During baking, the heat makes the gas bubbles expand, and the loaves become even lighter and puffier. Once the loaves are out of the oven and cooled, the tasty bread can be eaten.

WHEAT AS FOOD

The importance of wheat to the human diet is enormous. Wheat is one of the most nutritious foods in the world. A person could live on just whole wheat bread and water for a long time. Wheat grows without irrigation in dry climates where other food crops would fail. And so many different kinds of tasty foods that are easy to store can be made from wheat. It is no wonder that wheat is such a vital crop around the world.

The yeast mixture is added to the flour, along with eggs, salt, sugar, and oil.

The lumpy dough does not look promising.

The dough is kneaded.

After kneading, the dough is stretchy.

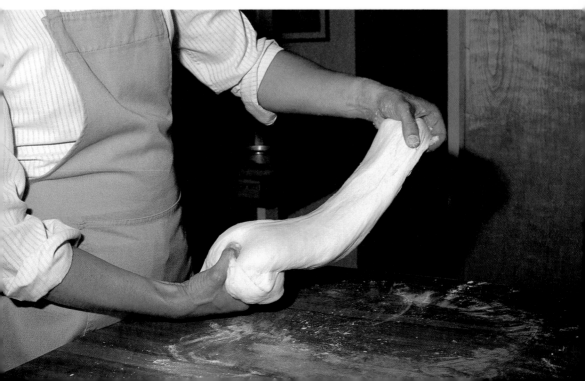

On the left, a ball of dough before rising. A ball of risen dough is on the right.

After rising, the loaf is brushed with egg to make it shiny, and sesame seeds are sprinkled on top. Then it is baked.

Loaves at a bakery in the oven.

The freshly baked bread in a bakery.

Golden fields of barley alternate with red wheat in Washington's hilly

wheat-growing country.

GLOSSARY

Bran—the outer covering of a wheat kernel.

Bulgur—cracked, steamed, and dried bits of wheat cooked like rice.

Cereals—grain crops. *See* Grain.

Chaff—bits of the head of grain other than the kernels.

Combine—a machine for harvesting grains that combines the tasks of reaping and threshing into one operation.

Disking—working the soil with a set of metal disks that cut deeply into the soil.

Durum wheat—a kind of very hard wheat used to make pasta.

Elevator—a building used to store grain.

Endosperm—the biggest part of the kernel, it provides food for the growing plant.

Erosion—the wearing away of soil by wind or water.

Fallow field—a field left without a crop so it can accumulate water and nutrients.

Germ—the part of the wheat kernel that would grow into a plant.

Gluten—protein found in wheat that gives bread dough stretchiness.

Grain—a seed of some plants, such as wheat and rye, belonging to the grass family.

Grain drill—a machine used to plant grains.

Hard wheat—wheat varieties with lots of gluten.

Harrow—machinery with many small teeth used to smooth the ground before planting.

Head—the mature spike of wheat seeds.

Herbicide—a chemical used to kill weeds.

Irrigate—to water a crop.

Kernel. *See* Grain.

No-till—a method of farming in which the soil is not turned; herbicides are used to kill the weeds instead.

Pasta—macaroni and noodle products.

Reaping—cutting a crop plant such as wheat.

Semolina—coarsely ground durum wheat used to make pasta products.

Soft wheat—wheat with little gluten.

Spike—the flower stalk of a grain crop such as wheat.

Spring wheat—varieties of wheat usually planted in the springtime.

Straw—dried stems and leaves left over after the grain has been removed.

Stubble—the short remains of cut grains left in the field after harvest.

Threshing—separating grain from stems and husks.

Unleavened bread—bread made without yeast.

Winter wheat—varieties of wheat planted in the fall and harvested the following year.

INDEX